# 1,000,000 Books

are available to read at

# Forgotten Books

www.ForgottenBooks.com

Read online
Download PDF
Purchase in print

ISBN 978-1-333-93649-5
PIBN 10563543

This book is a reproduction of an important historical work. Forgotten Books uses state-of-the-art technology to digitally reconstruct the work, preserving the original format whilst repairing imperfections present in the aged copy. In rare cases, an imperfection in the original, such as a blemish or missing page, may be replicated in our edition. We do, however, repair the vast majority of imperfections successfully; any imperfections that remain are intentionally left to preserve the state of such historical works.

Forgotten Books is a registered trademark of FB &c Ltd.
Copyright © 2018 FB &c Ltd.
FB &c Ltd, Dalton House, 60 Windsor Avenue, London, SW19 2RR.
Company number 08720141. Registered in England and Wales.

For support please visit www.forgottenbooks.com

# 1 MONTH OF FREE READING

at

www.ForgottenBooks.com

By purchasing this book you are eligible for one month membership to ForgottenBooks.com, giving you unlimited access to our entire collection of over 1,000,000 titles via our web site and mobile apps.

To claim your free month visit:
www.forgottenbooks.com/free563543

\* Offer is valid for 45 days from date of purchase. Terms and conditions apply.

**English**
**Français**
**Deutsche**
**Italiano**
**Español**
**Português**

# www.forgottenbooks.com

**Mythology** Photography **Fiction** Fishing Christianity **Art** Cooking Essays Buddhism Freemasonry Medicine **Biology** Music **Ancient Egypt** Evolution Carpentry Physics Dance Geology **Mathematics** Fitness Shakespeare **Folklore** Yoga Marketing **Confidence** Immortality Biographies Poetry **Psychology** Witchcraft Electronics Chemistry History **Law** Accounting **Philosophy** Anthropology Alchemy Drama Quantum Mechanics Atheism Sexual Health **Ancient History Entrepreneurship** Languages Sport Paleontology Needlework Islam **Metaphysics** Investment Archaeology Parenting Statistics Criminology **Motivational**

# A WHEATSTONE BRIDGE FOR RESISTANCE THERMOMETRY

By C. W. Waidner, H. C. Dickinson, E. F. Mueller, and D. R. Harper 3d

## CONTENTS

|  | Page |
|---|---|
| I. PRINCIPLES OF DESIGN. | |
| 1. Introduction | 571 |
| 2. Mercury links and shunt dials | 572 |
| 3. Improvements from an older instrument of same type | 572 |
| 4. General features of a thermometer bridge | 573 |
| II. DETAILS OF DESIGN AND CONSTRUCTION. | |
| 1. General description | 574 |
| 2. Sealed coils | 575 |
| 3. Mercury-cup contact blocks and links | 577 |
| 4. Total immersion in oil | 578 |
| 5. Link lifters | 578 |
| 6. Mechanical support and marble top | 579 |
| 7. Insulation resistance and equipotential shield | 579 |
| 8. Link and dial contact resistances | 580 |
| 9. Ratio reversing commutators | 580 |
| 10. Thermometer connectors | 581 |
| 11. Battery distributing switch and general arrangement of connections | 581 |
| 12. Thomson bridge connections | 583 |
| 13. Temperature control system | 584 |
| III. CALIBRATION. | |
| 1. Principles | 585 |
| 2. Manipulation and apparatus | 585 |
| 3. Special features of Thomson bridge calibration | 587 |
| IV. PERFORMANCE | 587 |
| V. SUMMARY | 590 |

## I. PRINCIPLES OF DESIGN

### 1. INTRODUCTION

By use of resistance thermometers, temperature measurements are frequently made to an accuracy of 0°.001 and in the measurement of small temperature changes, as in calorimetry, 0°.0001 or even less is sometimes desired. The former requires resistance measurements with an accuracy of about 1 in 300 000, while the latter, involving the measurement of a small change in a large quantity, requires a precision in each measurement of the order of 1 part in 3 000 000 or more. In nearly all cases in which such measurements have been made the electrical apparatus has been of special design.

## 2. MERCURY LINKS AND SHUNT DIALS

To attain the above accuracy in measurements of resistances of the magnitudes usual in resistance thermometry (1 ohm to 100 ohms), the errors due to contact resistances of plugs, switches, etc., must be minimized. One familiar method of doing this is by use of mercury-cup contacts for links connecting the parts of the circuit. This is fairly satisfactory for the larger valued decades of a Wheatstone bridge, but is impracticable for several reasons in the decades comprising small fractions of an ohm. An excellent device for these which fulfills the requirement of greatly diminishing the effect of dial contact resistances is constructed on the well-known principle of changing the resistance of a circuit by shunting a small resistance with a much larger variable one.[1] The contact resistance of the switch is here a part of the shunting branch, and only a small fraction of its total variation enters into the final result.

## 3. IMPROVEMENTS FROM AN OLDER INSTRUMENT OF THE SAME TYPE

About 12 years ago Messrs. Waidner and Wolff, of this Bureau, designed bridges embodying the above features. The coils of the main arm of one of these bridges were of the 5, 2, 2, 1 series from 50 ohms down to 0.01 ohm, and were connected by mercury contact links. The ten steps of 0.001 ohm each were secured by shunting a fixed resistance of 0.7200 ohms with a set of resistances

---

[1] Such a device for producing changes in resistance by small, even-valued, and equal steps was described in a paper communicated by Waidner and Dickinson, of this Bureau, to the American Physical Society in 1904. However, the printed abstract of this communication (footnote 2) is so very condensed as to contain only a vague suggestion of the arrangement used. Since this date the device has been described by White, Zeitschrift für Instrumentenkunde, 27, p. 211, 1907; Diesselhorst, Ibid, 28, p. 2, 1908, and White, Ibid, 34, p.112, 1914. Diesselhorst attributes it to White, who in turn mentions in his later paper its development several years ago by Waidner and Wolff at the Bureau of Standards.

extending from about 50 ohms to ∞, and of values such that the change introduced in the equivalent resistance of the divided circuit was just 0.001 ohm for each step through which the controlling dial switch was turned. A similar plan with different valued coils was used for the 0.0001 ohm and 0.00001 ohm decades. This bridge is shown in Fig. 1 and has been briefly described elsewhere.[2] It has given very satisfactory service for over 10 years, but in the course of this time experience has suggested a number of improvements, mostly relating to convenience of operation. The more important are:

(a) Diminution or elimination of the seasonal changes of the resistance coils due to variations in atmospheric humidity. The interval between calibrations of the bridge could then be greatly extended without impairing the accuracy of work done with it.

(b) Reduction of thermoelectromotive forces occurring at the link and dial contacts when changing a setting.

(c) Simplification of the manipulation to permit of greater ease in following rapid changes of the resistances measured with the bridge.

### 4. GENERAL FEATURES OF A THERMOMETER BRIDGE

The following principles, dictated partly by the experience gained in the use of the older bridge, served as a basis for the design of the one here described:

(a) The coils should be hermetically sealed to protect them from the influence of atmospheric humidity.

(b) The whole bridge proper—that is, connecting switches and links as well as coils—should be immersed in the thermostat.

(c) The operating mechanism necessitated by requirement (b) should be rigidly aligned, whence all the materials of construction should be as permanent as possible. (One application of this principle was the substitution of marble for hard rubber.)

(d) The variable resistance arm should consist of six decades.

(e) The construction of the contacts and other resistance arrangements should be such that readings could be made to an accuracy of 2 or 3 per cent of one step of the last decade.

---

[2] Waidner and Dickinson: Physical Review, 19, p. 51, 1904. Schematic diagrams of the instrument with a few descriptive lines are given in this Bulletin, 3, p. 646, 1907; and 6, p. 153, 1909.

(*f*) The arrangement should be such that measurements be possible by the Siemens or the Callendar method of connecting a thermometer, or by the use of the Thomson double bridge.

(*g*) The arrangement should be such as to permit of calibration easily and to an accuracy indicated in paragraph (*e*).

## II. DETAILS OF DESIGN AND CONSTRUCTION

### 1. GENERAL DESCRIPTION

The assembled bridge is shown in Fig. 2. The general construction is shown by Figs. 5 and 7. Upon a marble plate are mounted all the contact blocks and switches and from it also the coils are supported. All the metal parts are raised about 3 mm from the plate by small hard rubber blocks. The marble plate is mounted on four steel posts which are supported by a brass casting, the whole forming a rigid protective and supporting structure. This structure carries an auxiliary top, also of marble, to which are fastened the link lifters, handles of the dials, and other manipulating devices, but no portion of the electrical circuits. When the bridge is immersed in the thermostat oil bath the upper surface of the oil comes between the two marble plates, about a centimeter higher than the top of the contact blocks.

Diagrams showing the electrical features of the bridge are given in Figs. 11, 12, and 13, discussed in detail later. The principal parts of the bridge as built are:

(*a*) A pair of 100-ohm and a pair of 1000-ohm ratio coils, either pair of which can be introduced into the circuit at will and interchanged to eliminate error due to inequality of the two coils.

(*b*) A variable resistance arm composed of six decades, the first three being series of coils on the 5, 2, 2, 1 plan, ranging in values from 50 ohms to 0.1 ohm and connected to copper mercury-cup contact blocks provided with amalgamated copper links, and the last three being obtained by shunting resistances of 2.2 ohms, 0.34 ohms, and 0.071 ohms, respectively, with appropriate shunts (values marked on Fig. 12) to give steps of 0.01 ohm, 0.001 ohm, and 0.0001 ohm.

(*c*) A compensating coil of about 2.5 ohms (more exactly the value of the last three decades described in (*b*) when the dials are

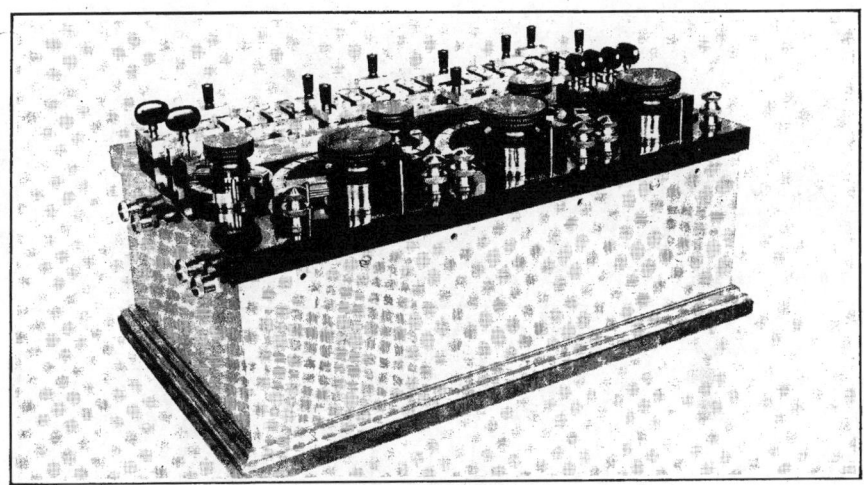

FIG. 1.—*Special Wheatstone bridge for resistance thermometry. Designed and constructed in 1903*

FIG. 2.—*Special Wheatstone bridge for resistance thermometry. Designed and constructed in 1911*

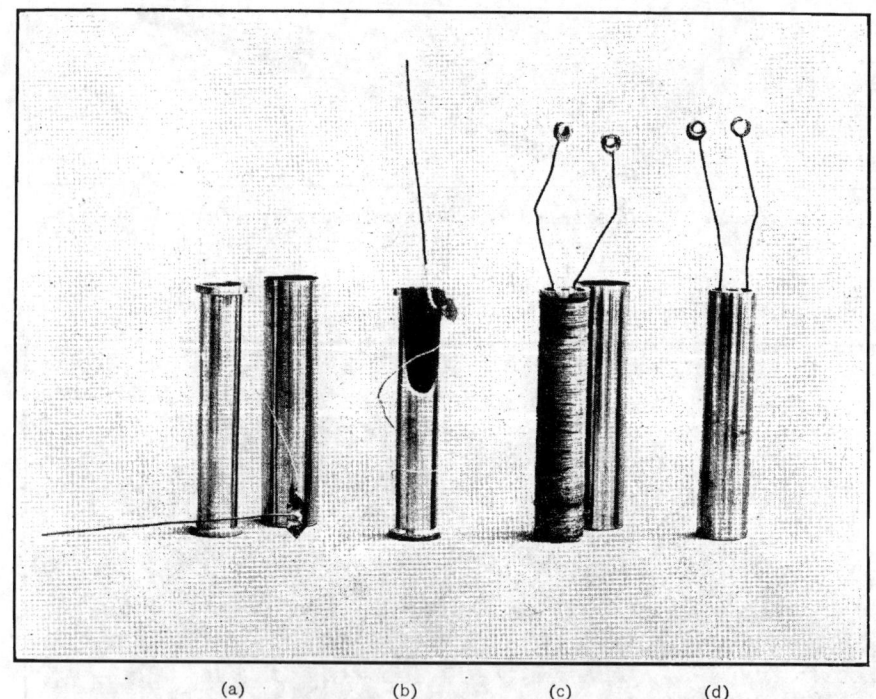

FIG. 3.—*Hermetically sealed coils for Wheatstone bridge. Details of construction*

set on zero) in the arm of the bridge in which the thermometer is to be connected, which coil makes the bridge direct reading, and permits of measuring resistances less than 2.5 ohms.

(d) Three coils and a slide wire so arranged as to provide auxiliary ratio arms which in combination with the simple bridge form a Thomson double bridge.

(e) A battery distributing switch.

(f) A galvanometer switch to change the connecting points of the galvanometer appropriately for the Thomson bridge method.

(g) Battery and galvanometer reversing switches.

(h) Sensibility switch, changing the emf applied to the bridge.

## 2. SEALED COILS

The investigations of Rosa, Dorsey, and Babcock [3] showed the desirability of protecting resistance coils from the influence of atmospheric humidity, and developed a method [4] of doing so for individual resistance standards, but no sealed coil suitable for use in resistance boxes has previously been developed,[5] and the plan which has been suggested of sealing the entire box did not appear feasible. A form of sealed coil suggested by the heating coils which have been used for a number of years in the various calorimeters in the laboratories of the Bureau was developed and adopted.

The construction of the coils is shown in Fig. 3. At (a) is a group of some of the parts. The flanged metal spool will slip into the tube just to the right of it, and when the ends are soldered, there results the hermetically sealed annular space in which the wire is wound, as shown at (c). Some provision is necessary for bringing out the terminals of this wire without impairing the insulation or permitting leakage. Several devices were proposed and we are indebted to Dr. F. Wenner, of the Bureau, for the one adopted. The details are shown at (a) and (b), Fig. 3. To the point of a thin strip of sheet copper cut in the shape shown was hard soldered the

---

[3] Rosa and Dorsey, this Bulletin, 3, p. 553, 1907; Rosa and Babcock, this Bulletin, 4, p. 121, 1907 (Scientific Paper No. 73).

[4] Rosa, this Bulletin, 5, p. 413, 1908 (Scientific Paper No. 107).

[5] A sealed coil is described in Zeitschrift für Instrumentenkunde, 33, p. 126, 1913, as a modification of a form developed by the Bureau of Standards. The reference is probably to the coil described below which was designed in 1910. (See p. 587.)

end of the insulated manganin wire of suitable size for the coil winding, and at the center of the same copper strip, perpendicular to its plane, was hard soldered a stout piece of copper wire. After providing suitable insulation, this was passed, as shown in Fig. 3 (*b*), through a hole drilled in the metal spool. The insulation was secured in the following way: The copper strip was covered with two layers of silk gauze, attached by painting with alcoholic solution of shellac, and the wire was wrapped beyond the bend with silk thread. When the shellac solution had dried, dry shellac was melted in, thoroughly impregnating the silk. The spool also was wrapped with shellacked silk gauze and the part to be covered by the copper strip was impregnated with melted shellac. While hot the terminal was pressed tight against the spool and permanently tied in place with silk thread. The freezing of the melted shellac formed an air-tight seal over and around the hole drilled through the spool. Melted shellac flowed into this, binding the lead wire immovably and providing excellent insulation of the latter. The use of dry shellac melted into the silk seems necessary, the use of shellac solution alone failing to give a good seal.

In Fig. 3 the terminal strip is shown bare at (*a*) and covered with silk at (*b*), the spool being bare in both cases. At (*c*) is shown a coil fully wound and wrapped with silk after winding. At (*d*) is shown a coil with cover in place, finished and ready for use.

Adjustment of the coils proceeded as follows: Very nearly the correct value was attained by using the proper length of wire in the coil winding. A closer adjustment was made, after the coil was annealed, by clipping off suitable lengths of the wire at the lower end of the coil, where the wires double back in the noninductive winding, and hard soldering these ends together. The final adjustment was made upon the leads after the coil had been mechanically completed and mounted in place. It is desirable that the external leads of the coils be of manganin, and accordingly there is considerable latitude for such adjustment. The more flexible copper is, however, advantageous at the bend, so that a copper lead should be used as described above, and then cut off just beyond the bend and a suitable size piece of manganin wire hard soldered on.

The construction described produced a coil very similar in size, shape, and general appearance to the usual form of open coil, except for the smooth brass tube instead of the furrowed outer surface and for the leads coming out from the inner spool. The carrying capacity when immersed in oil is not quite so great as for a corresponding uncovered coil, but the difference is not sufficient to be of importance. The heating for given current is about one and one-half times the heating of the open coil.

No great advantage results from the hermetical sealing of very low resistances, and in the present bridge the construction described was employed only for coils whose resistence exceeded 0.3 ohm.

### 3. MERCURY-CUP CONTACT BLOCKS AND LINKS

The lugs forming the terminals of the coils shown in Fig. 3 were soldered to copper posts projecting down from blocks containing mercury cups, the construction being shown in Fig. 4. When not wanted in the circuit the coil is short circuited by a link, and upon raising this the resistance of the coil is substituted for that of the copper link plus the two mercury contacts between the cups and the studs on the link. Upon the constancy of these contacts depends the reliability of the bridge and the main factor in securing the desired constancy is proper construction to secure plane surfaces and the bringing into one plane of all the four surfaces concerned. The cups were purposely made shallow, about 2 mm. deep, so as to be readily cleanable.

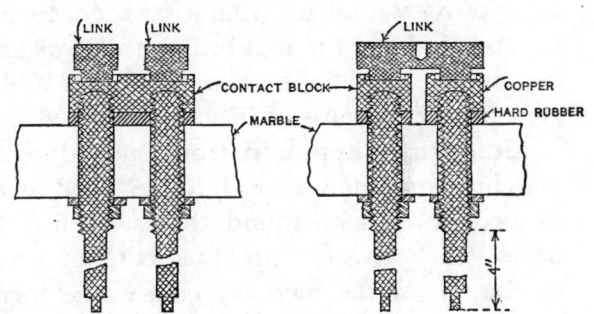

FIG. 4.—*Mercury-cup contact blocks and links*

The use of a rod for each coil terminal, namely, two for each contact block as shown in Fig. 4, makes the settings single valued. This is not true for a construction employing one rod for each block and attaching two coils to it at its lower end.

### 4. TOTAL IMMERSION IN OIL

The requirements as to accuracy make it necessary to submerge the coils of such a bridge in a thermostat oil bath, and to arrange for proper working of a convenient thermostat, the temperature at which it is set must be somewhat in excess of that of the room. If the links, etc., emerge from the oil bath into the air of the room, the temperature differences, even though small, are usually sufficient to be the cause of troublesome thermal electromotive forces. Hence, in the present bridge all parts of the electric circuits were placed within the thermostat bath. The design adopted is sufficiently well shown by Figs. 5 and 7, in connection with what has already been said under "General description," page 574.

### 5. LINK LIFTERS

The immersion of the links and dials in oil rendered necessary some provision for manipulating them through a cover. A second top, 6 cm above the real bridge top, was arranged as a mechanical keyboard. The links are raised from and lowered into the mercury cups by means of lifters illustrated in Fig. 6.

Although it appeared from such data as were available that mercury contacts under oil are as good as or better than the same in air, it was soon found that with link lifters which set a link down gently a very unreliable contact was obtained. The action of the oil on the mercury causes the formation of a film which must be broken before proper contact is secured. Accordingly, the lifter had to be such as to communicate to the link a motion which would break the film, and also would leave the link entirely free to rest on its own base plane when all the way down. The requirements are satisfied by the form of lifter shown in Fig. 6. A yoke fitting freely in slots in the ends of the link, together with a central pin resting in a cup in the link, permit of applying the necessary vertical pressure and at the same time "scrubbing" the contact by a small rotation of the link communicated by means of the knob forming the handle of the lifter. The extent of this rotation is limited by pins in the bushing of the lifter so that the link can never be turned far enough to fail to seat in the mercury cups of the contact blocks. When a link is lowered into place

FIG. 5.—*Showing the actual bridge top with its mercury-cup contact blocks and dial switches completely immersed in oil. The cover bearing the manipulating devices is shown at the back*

FIG. 6.—*Link and link lifter*

FIG. 7.—*Bridge removed from oil bath thermostat; to show rigid structural framework, mounting of coils both sealed and open, and also the details of the manipulating devices. To show the latter, the top is poised at an angle*

and the hand removed from the lifter the link is entirely free from constraint except for support by its own studs.

The operation of the dials from the top plate of the bridge is illustrated by Fig. 7. An ordinary hard-rubber dial handle with the usual clicking device was mounted on the plate, and its shaft extending downward with a cross key at the bottom engages a cross slot in the top of the dial mechanism.

### 6. MECHANICAL SUPPORT AND MARBLE TOP

The manipulating devices for the dials and links required careful and permanent registering with the parts below. This was secured by the use of the framework briefly described above (p. 574) and sufficiently well shown in Fig. 7 to obviate the necessity of detailed description. This frame supported the bridge in the proper position in the oil-bath thermostat.

To insure that alignment once attained would be permanent and not be destroyed by warping, marble was selected in preference to hard rubber as the material for the main plate of the bridge. The upper plate carrying the manipulating handles is also of marble. The only serious disadvantage thus incurred was a considerable increase in the mechanical difficulties of construction.

### 7. INSULATION RESISTANCE AND EQUIPOTENTIAL SHIELD

Since marble may be defective in insulation qualities because of veins of conducting material, the plate used was thoroughly tested for insulation. After drilling, each hole was stoppered and filled with mercury and a test made with a voltage of about 300. No insulation resistance of less than 30 000 megohms was found between any two points where metal parts were to be placed. The plate was of white marble of good clear appearance.

The presence of high potentials (relative to those in Wheatstone bridge measurements) in almost every physical laboratory makes it necessary that all electrical apparatus intended for very precise measurements shall be adequately protected from chance leakage of such stray voltage into the measuring circuit. Because of the possibility of a low surface insulation resistance, even when the best of insulating material is employed, it is desirable to screen such apparatus with a suitable metal network to equalize any differences of potential which may result in such leakage. The screen for the present apparatus was constructed by joining electrically the copper tank containing the oil in which the bridge was immersed, metal plates in the feet carrying the board upon which the assembled apparatus was mounted, and a metal plate beneath the galvanometer.

### 8. LINK AND DIAL CONTACT RESISTANCES

The total resistance of a link and its two mercury contacts is about 13 microhms and is constant to about 1 microhm when operated by the form of lifter described in section 5 (p. 578) with a scrubbing motion to break up the surface film on the mercury.

The variations of resistance of the contacts of the shunt dials introduce no error greater than a microhm. By referring to Fig. 12 it will be seen that in the worst case, namely, with the largest step dial on the zero setting, the figures are as follows: 2.2 ohms is shunted by 46.2 ohms, making a joint resistance of 2.1 ohms. To change this resistance by 1 microhm the shunt of 46.2 ohms must be increased or decreased by 0.0005 ohm. With well-made dial contacts under oil this variation in contact resistance is not to be expected and has not been found.

A convenient way of stating the result deduced above, namely, that a variation of 0.0005 ohm in the shunt circuit corresponds to a change of 0.000 001 ohm in the bridge circuit is to say that the effects of variation of the contact resistances are reduced 500 times. So stated, the effect is reduced 1000 and 5000 times for the zero positions of the middle and smallest dials, respectively, while on the settings at nine the figures are 50 000, 100 000, and 500 000, respectively, for the three dials, with intermediate values for the intermediate settings. It is therefore evident that even if contact resistances were liable to much greater variation than 0.0005 ohm no serious errors would be introduced.

### 9. RATIO REVERSING COMMUTATORS

The ratio coils are connected according to the plan shown in diagram in Fig. 11, which permits of interchanging the ratio arms of the bridge by rotating a four-point commutator through 90°. The resistance of the commutator contacts enters directly into the ratio circuit, and mercury cup contacts were constructed with the same care as for the other arms of the bridge. The details are shown in Fig. 8, which illustrates how the links are raised and

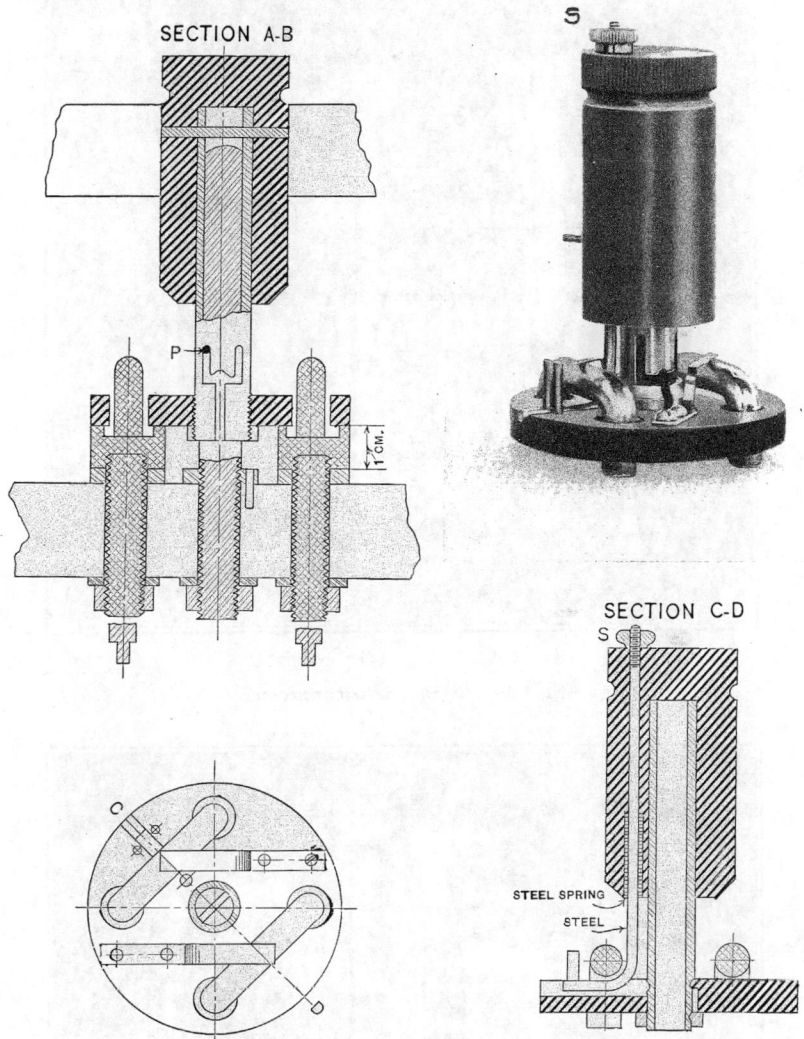

FIG. 8.—*Ratio interchanging commutator*

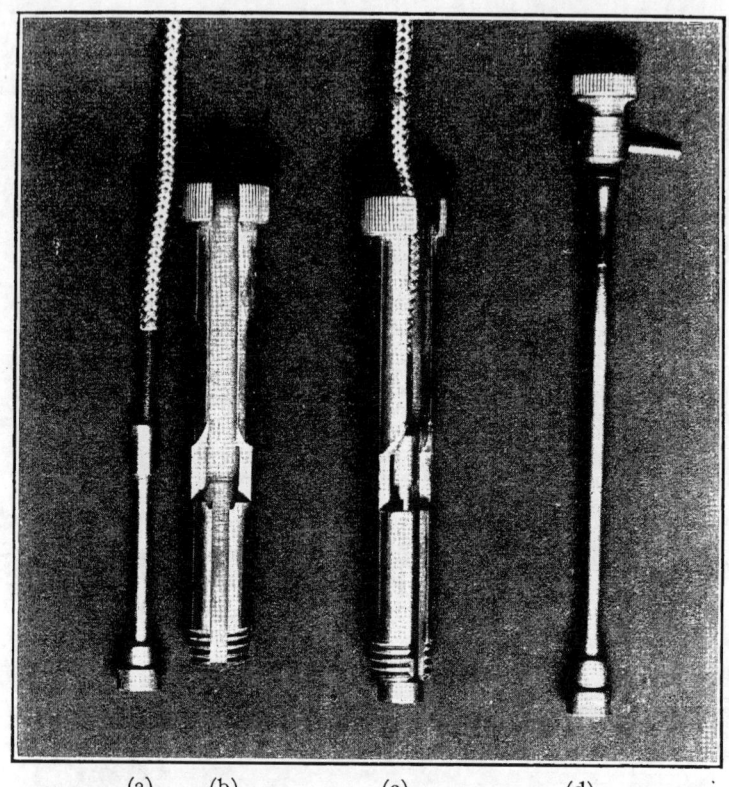

(a) (b) (c) (d)

FIG. 9.—*Thermometer connectors*

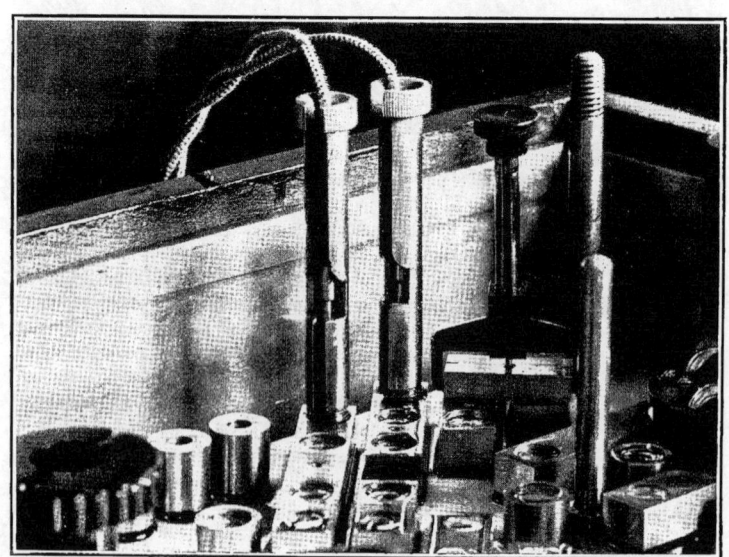

moved, and also are free to rest on their base plane when lowered into the cups.

In Fig. 11, for the sake of simplicity, but one pair of ratio coils is indicated. There are two pairs, one each of 100 ohm and 1000 ohm coils, connected as shown in Fig. 12. When using an equal-arm bridge connection, one of the two commutators is raised entirely clear of the mercury cups and suspended by the pin (P), Fig. 8. (See also Fig. 2, which shows one commutator raised and one lowered.) When using the 10:1 or 1:10 ratio, both commutator handles are lowered into position and one link of each raised by a screw (S), Fig. 8, so as to be out of contact with its cups.

## 10. THERMOMETER CONNECTORS

To secure the advantages of the mercury contacts for the bridge coils, it is desirable to have equally good contacts in connecting the thermometer to the bridge. This is done by means of connectors illustrated in Fig. 9. The flexible wires represent the thermometer leads. To these are soldered the special terminals (a) which fit in the mercury cups and are held securely in place by long nuts (b). At (d) is shown a similar terminal with a binding post top for making connections when a soldered connection is unnecessary or inconvenient. In Fig. 10 is shown the way in which this type of connector attaches to the bridge, the upper marble plate having been removed so as not to obstruct the view.

## 11. BATTERY DISTRIBUTING SWITCH AND GENERAL ARRANGEMENT OF CONNECTIONS

One terminal of the battery is connected at a fixed point in the bridge, namely, between the two ratio arms; the other terminal is left for connection at whatever point of the circuits it is desired to make the dividing point between the two remaining arms of the bridge. This depends entirely upon the use of the bridge, being different for the Siemens and Callendar pattern resistance thermometers, for measurements by means of the Thomson double bridge, for calibration, etc.

A complete diagram of connections is given in Fig. 12, and the general plan, with details omitted, is shown somewhat more clearly by Fig. 11. The points of the battery distributing switch (Fig. 12) function as follows: The point A is connected to a binding post used for the battery connection of a Siemens type thermometer and for many general resistance measurements; B is used in conjunction with a Callendar thermometer, the main coil of which is connected in the gap 7-8 and the compensating coil at 5-6; C replaces A when both connecting leads of an external resistance are to be thrown

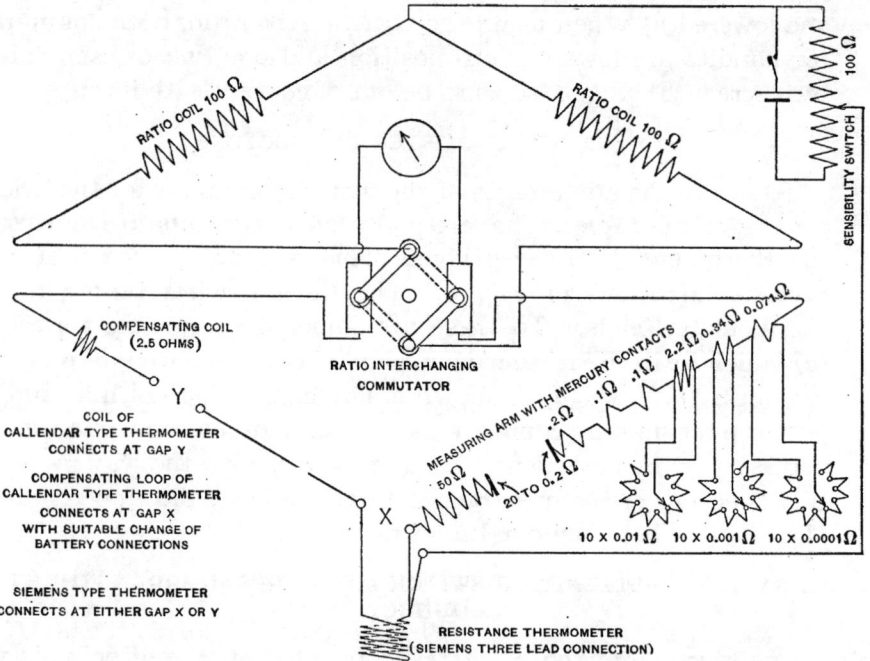

FIG. 11.—*Diagrammatic representation of circuits. Connections as a simple bridge; switches for battery reversal, galvanometer reversal, etc., omitted.*

into one arm of the bridge instead of so as to compensate each other; D is for checking the constancy of the mercury contact resistances [6]; E is for the Thomson bridge.

---

[6] The method by which the link resistances are determined is as follows: With all the links between C and D lowered and suitable connectors in the gaps 5-8, the bridge is balanced with the battery distributor set at C and again at D. In the first instance the line of links and contacts is in the same arm of the bridge as the shunt dials; in the second instance it is in the opposite arm. Twice the resistance of this line, i. e., 26 links, is the difference in the two settings. If this be found too large, a more extended calibration is in order for the location of the faulty contact. By this arrangement—that is, having C and D permanently connected as they are—it is but a moment's work to check up the contacts, and it is therefore much more likely to be done frequently than if external connections had to be made each time.

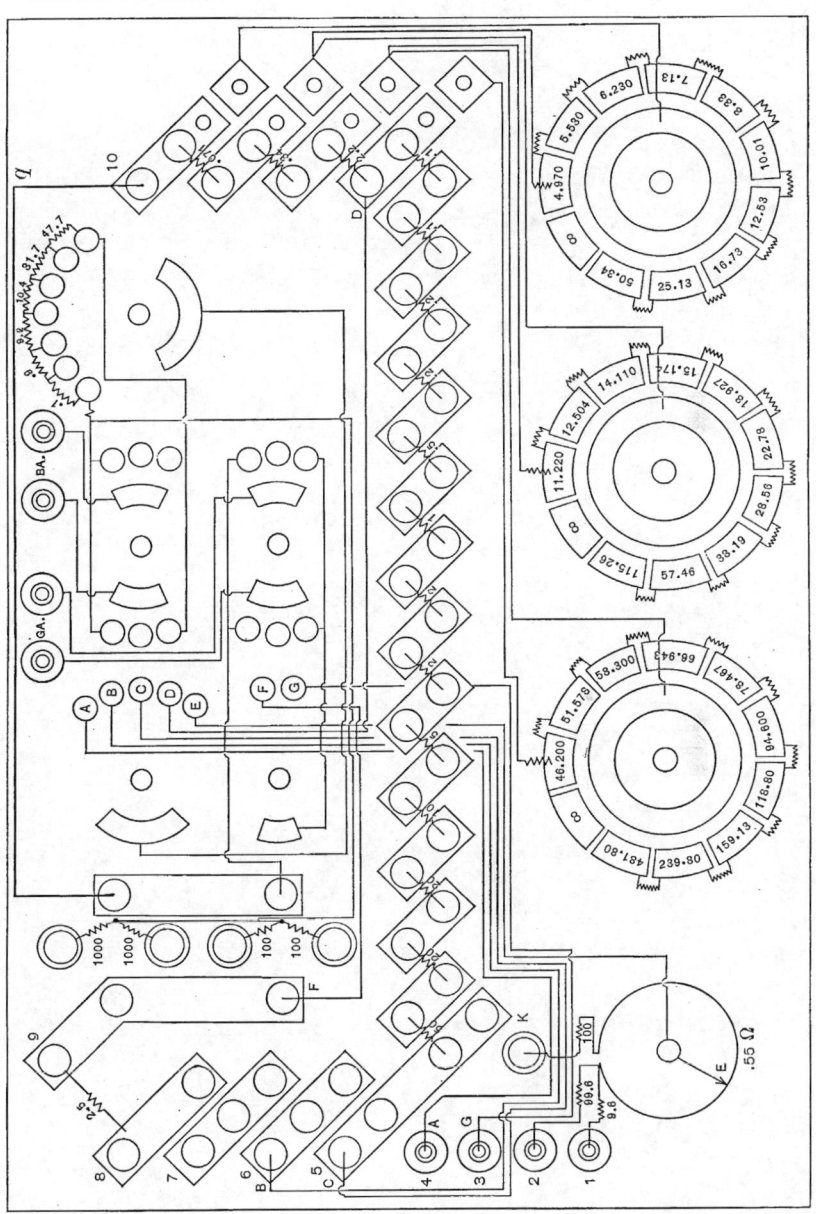

FIG. 12.—*Actual connections of bridge and arrangement of bridge top*

The other auxiliary switches require little explanation. The battery and galvanometer reversals are accomplished by four-point commutating switches, which include an "off" position. The sensitivity switch is in the battery line and operates by tapping off a given fraction of the applied emf. The first point corresponds to about a hundred-thousandth full sensitivity, so that a totally unbalanced bridge gives less than full-scale deflection of the galvanometer. This is very useful in determining the magnitude of an entirely unknown resistance. Successive steps increase the sensitivity until the full battery emf is applied to the bridge.

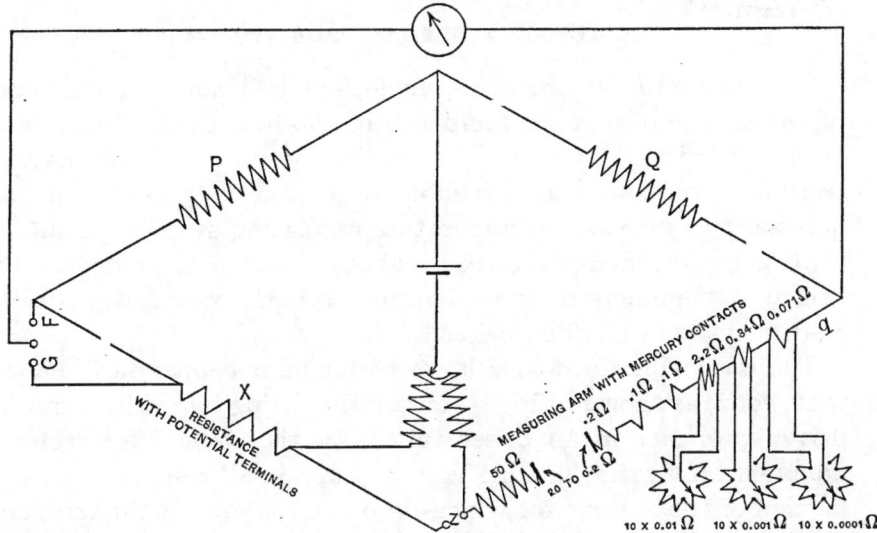

Fig. 13.—*Diagrammatic representation of circuits when connected as a Thomson double bridge.*

### 12. THOMSON BRIDGE CONNECTIONS

The diagrammatic Thomson bridge [7] is shown in Fig. 13, and with the aid of this the actual connections may be traced out in Fig. 12, if desired. By depressing the link at K the auxiliary ratio is connected into the simple bridge at one end of the variable arm, corresponding to the arrangement shown in Fig. 13. The

---

[7] The theory of the Thomson bridge can not be discussed here. The original paper on the method was published by Sir William Thomson, Philosophical Magazine, 24, p. 149; 1862. The subject is fully treated, with a comprehensive bibliography, in a paper by Wenner, "The Four-terminal Conductor and the Thomson Bridge," this Bulletin, 8, p. 580; 1912 (Scientific Paper No. 181).

gap Z (Fig. 13) is provided by the link 5–6, Fig. 12, and the auxiliary ratio and its slide wire are in the extreme lower left corner of Fig. 12. The choice of ratios, equality or 10 to 1, is clearly shown by Fig. 12 and, for the sake of simplicity, is omitted from Fig. 13.

If it be necessary to use ratio coils of high resistance, the usual method of connecting a Thomson bridge seriously limits the sensitivity attainable,[8] so the connections shown in the figures were adopted, namely, with the resistance of the main and auxiliary ratios in the battery circuit and not directly in the galvanometer circuit, a subject which has been somewhat more fully discussed by Wenner.[9]

### 13. TEMPERATURE-CONTROL SYSTEM

The maximum temperature coefficient of change of resistence of any of the coils, at the temperature at which the bridge is used, is of the order of 15 parts per million per degree C, so that for resistance measurements reliable to 1 part in 3 000 000 it was necessary to provide a temperature-regulating system capable of controlling the temperature to about 0.02°. Accordingly, the bridge was immersed in a thermostat bath provided with the control system described below.

The oil is contained in a large rectangular copper tank lagged with wood, as shown in Fig. 5, circulation being forced by a motor-driven propeller in an offset tube. In the same offset tube is mounted an electric heating coil of "advance" resistance wire, a portion of the current for which is controlled so that the temperature of the bath remains constant within about 0.03° and that of the coils within 0.01°. The controlling device is a relay operated by a mercury make-and-break, the motion of which is due to expansion and contraction of the liquid in a large bulb distributed over the bottom surface of the oil bath. The relay, battery, switches, fuses, etc., are all assembled on a board at the end of the table carrying the bridge, as shown in Fig. 2.

The temperature of 30° C was selected for the operation of the bridge, and the apparatus is arranged to heat up rapidly to this temperature by using a 120-volt supply circuit, and then by the

---

[8] W. Jaeger und H. von Steinwehr: Annalen der Physik, (4), **43**, p. 1165; 1914.
[9] Wenner, loc. cit., p. 594.

throw of a multiple-blade switch to regulate on 22 volts. A bell is connected so as to give warning when, on the rapid heating, the temperature of 30° has been reached, or if, on regulating, the relay fails to operate.

## III. CALIBRATION

### 1. PRINCIPLES

The general principle of calibration is to compare directly the various coils and series of coils in the bridge, much as weights are compared, i. e., 50 with $20_1 + 20_2 + 10$, one such comparison being made with an external standard resistance, usually 100 ohms, so that all the results are expressible in international ohms. From these comparisons enough equations are obtained to solve for the correction to each coil.

In making such comparisons the differences between the coils are read from the small decades of the bridge, i. e., the dial shunts, and these must either be calibrated previously or else the assumption made that they are correct. The latter procedure is by far the more convenient, and if upon working down from large to small decades it should appear that the corrections to the latter were appreciable, the differences read in terms of them could then be corrected, and the calibration recomputed accordingly. No difficulty was experienced in adjusting these decades to values where the corrections are negligible, and it is not to be expected that they will change by any significant amount.

### 2. MANIPULATION AND APPARATUS

Fig. 11 will serve as the reference figure. Closing the gaps X, Y, with appropriate links, the dividing point between the two arms of the bridge, namely, the point at which the battery circuit is attached, is moved along for successive steps so as to secure the comparisons 50 vs. $\Sigma 50$, $20_1$ vs. $20_2$, $20_1$ vs. $\Sigma 20$, $20_2$ vs. $\Sigma 20$ 10 vs. $\Sigma 10$, etc., down to $0.1_2$ vs. 0 to 10 of the first dial. Such a comparison is made by balancing the bridge twice, once with the appropriate coils in the bridge arms and once with these short-circuited by their links. Balance of the bridge may always be obtained if a link of sufficient resistance be connected in X or Y, because the compensating coil balances approximately the shunt-dial decades when set in their zero positions. Such a calibrating link is a small loop of copper or manganin, according to circumstances. Two of them are shown at (c), Fig. 14, together with a traveling plug used for the movable battery connection. This plug fits in a small hole at the center of each contact block (see Fig. 5), a row of small holes just over these in the cover plate being provided and stoppered with hard-rubber plugs when not so used.

The comparison of the $50 + 20_1 + 20_2 + 10$ of the bridge with a certified 100-ohm standard for the purpose of expressing all the results in international ohms is accomplished by connecting the standard in the gap X (Fig. 11) just as a thermometer would be connected. A sealed standard of the usual form is used, being shown at (b), Fig. 14, with a convenient stand, short-circuiting link, and connectors to fit the bridge.

Calibration of the dial decades is possible by at least two methods. One of these is to measure the actual values of the shunt coils and from these deduce the corrections for the bridge arm. As this method requires an auxiliary bridge adapted to the measurement of odd-valued resistances and involves rather tedious computations, another method has been deemed more convenient. This is to compare the successive steps of a dial with each other by comparing each in turn to the total interval of the next smaller dial, the steps of the last dial being intercompared by observing the galvanometer deflection produced by changing the setting one step. The method obviously requires that the dial whose total interval furnishes the comparison unit shall be set at 0 and 10, respectively, when the two balances of the bridge are secured. To accomplish this, a small variable resistance is connected in the gap X and adjusted to a value which will give the balances with the dials in the desired positions.

A variable resistance used for this purpose is shown at (a), Fig. 14. A slide wire is mounted upon a cylinder at the center of which is the bearing of an arm carrying the movable contact clip connected by a flexible lead to a binding post in the battery circuit of the bridge. The resistance of the sliding contact is therefore not in a measuring arm of the bridge. The slide wire may be shunted by suitable resistances to give it different effective values, and since

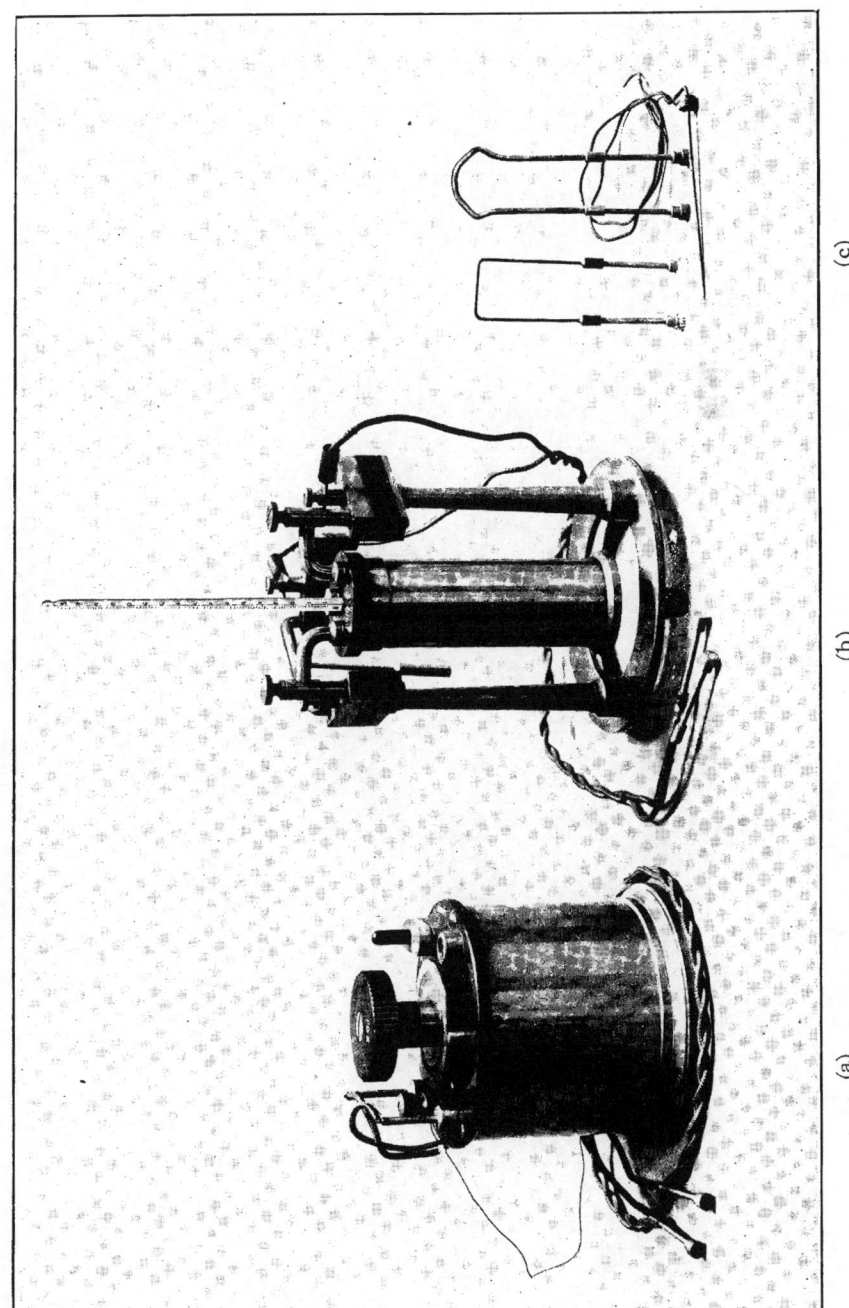

FIG. 14.—*Auxiliary apparatus for calibrating bridge*

the contacts for the shunts are directly in the bridge arms they are provided with mercury-cup connections.

Methods of connecting can be arranged so as to calibrate the separate parts of the shunt decades, i. e., the 2.2 ohm, 0.34 ohm, and 0.071 ohm coils, and the various connecting resistances, so that the bridge can be completely self-calibrated, a known outside resistance being needed only for the reduction from box units to international ohms.

### 3. SPECIAL CALIBRATION FOR THOMSON BRIDGE

For use of the apparatus as a simple Wheatstone bridge, the only calibration necessary is the determination of the substitution values of the coils in the main variable arm, the value of the steps in each dial, and the correction for inequality of ratio arms. The latter can be easily determined during the progress of any measurement. The magnitudes of the link resistances, connections, etc., are of no consequence so long as they remain constant. They are all gathered into one term and eliminated from a measurement by taking a "zero balance" of the bridge with a short circuiting link across the gap where the thermometer or other unknown resistance is connected.

For measurements with the Thomson bridge connection the same simple process can not be followed. The 2.5 compensating coil can not be used, no single step corresponding to the "zero balance" is feasible, and the absolute value of the resistance of each portion of the variable arm of the bridge must be known. Besides the series of coils 50 ohms to 0.1 ohm and their links, there are the actual values of the three shunted coils and the connecting bar, $q$, of Figs. 12 and 13.

## IV. PERFORMANCE

The instrument was delivered to the Bureau in January 1911. The coils in it at that time were found to be imperfectly sealed, and the final set was not installed until April 1912, so that data pertaining to the behavior of the sealed coils extend over an interval of about two years. The experience with the preliminary coils during their year of service was, however, not without value.

Certain coils of the first set had failed in their insulation resistance and so were replaced by ordinary open coils; the rest were sound electrically but leaked mechanically. By the alternate heating and cooling, such a coil soon becomes filled with oil in the spaces around the winding, so that in this respect it is not to be distinguished from an open coil. However, this oil is closely confined and communicates with the larger mass of oil through very small holes only, so that diffusion of the moisture content of the oil as the seasons change is very greatly hindered, and it might be expected that seasonal variations of resistance of such a coil would be quite small. The experience with the preliminary set of some open and some "partially sealed" coils proved this to be the case. The latter were found to be very much less subject to seasonal variation than the former, and were in fact much more constant than was anticipated.

The performance of the final set of sealed coils in the measuring arm of the bridge since April 1912, is shown by Fig. 15. There is no evidence of variation related periodically to a calendar year. The standard of reference is the international ohm as given by the mean value of 10 wire standards of the Bureau, the medium for the comparisons being a 100-ohm sealed standard as already described (p. 586; also Fig. 14). Except for the 10-ohm and 5-ohm coils, the total variation of any coil with respect to the standard unit is but a trifle more than 1 part in 100 000. In view of the small magnitude of this figure, combined with the fact that the curves all show the same general peaks and depressions, the question seems pertinent as to whether the absolute variation is not as likely to be in the standard of reference as in the bridge coils.

The behavior of the mercury contacts has been satisfactory. The formation of a troublesome surface film has already been mentioned (p. 578), but after the difficulties due to this were overcome no others remained. The average resistance of a link and its two contacts is 13.6 microhms (measured as indicated on p. 582). Of this, a resistance of at least 6 microhms is in the copper, whence the order of magnitude of the resistance of each mercury contact is 4 microhms. These have been found constant within 1 microhm except when the cups were extremely dirty.

The bridge has amply fulfilled expectations with respect to flexibility and accuracy. It has proven entirely satisfactory in meeting the requirements of a general extra-precise thermometric bridge. The principal objection which users have made against it is that the rather considerable size limits portability and requires an undue proportion of the space in a complex set-up where one operator is required to be within reach of a number of instru-

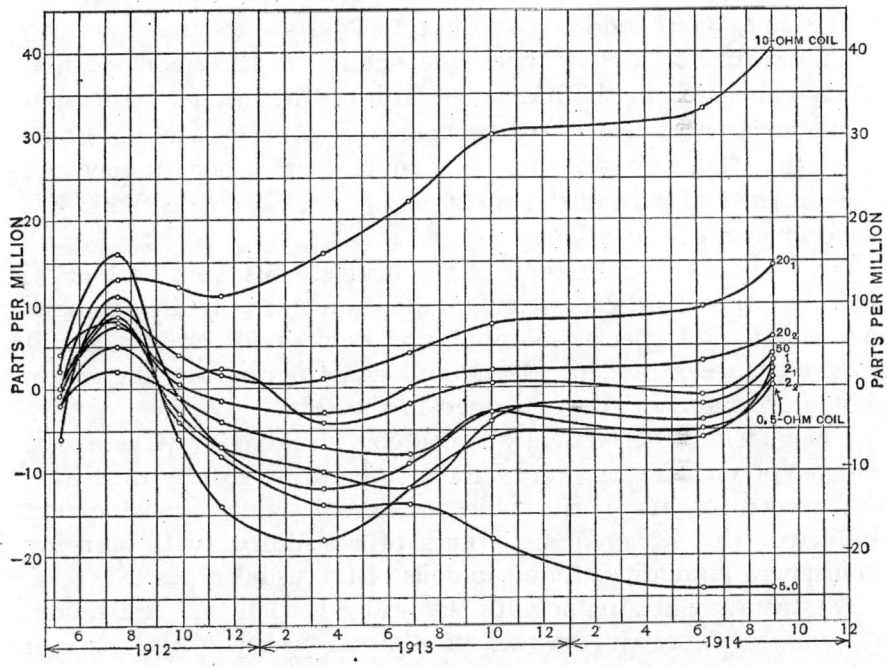

FIG. 15.—*Corrections to the sealed coils (in international ohms)*

ments. For a bridge of the mercury link type with the flexibility and accuracy possessed by this instrument it is not easy to see how the size can be materially reduced, but bridges of other types which are much more compact and perhaps somewhat more convenient are now being developed by the Bureau.

The Bureau is indebted to the Leeds & Northrup Co., who undertook the construction of the apparatus, for the care and skill with which many of the details of the designing as well as the construction were carried out.

## V. SUMMARY

The Wheatstone bridge described in this paper was designed with especial reference to flexibility of use in measurements with resistance thermometers. The bridge is adapted to use with either the Siemens type or Callendar type of resistance thermometer, or with the potential terminal type of thermometer by the use of the Thomson double-bridge method. The instrument is also arranged so that it may be completely self-calibrated.

The 0.01, 0.001, and 0.0001 ohm decades are secured by varying, by means of dial switches, the shunts on three coils permanently connected in the measuring arm of the bridge. The sum of the resistances which are permanently connected is 2.5 ohms when the dials are set on zero, so that in order to measure resistances smaller than this a coil of 2.5 ohms is connected in the adjacent arm of the bridge.

The entire electrical circuit of the bridge, coils, contact blocks, switches, and connectors are totally immersed in an oil bath thermostat, and special manipulating devices for the links and dials, etc., are provided. Details of construction are shown by photographs and briefly explained in the text.

A new form of hermetically sealed coil, suitable for Wheatstone bridges, potentiometers, and similar apparatus, is fully described and record of its performance reviewed. Such construction eliminates the seasonal variations of resistance (with varying atmospheric humidity) found in coils of the usual types.

The accuracy attainable with the bridge is such that resistances of 1 ohm or more can be measured to an accuracy of 1 part in 300 000 in terms of the unit in which the calibration is expressed. This corresponds to an accuracy of about $0°.001$ for measurements with the platinum resistance thermometer. Low resistances, the accuracy of measurement of which is limited by variations in contact resistances, may be measured to about three millionths of an ohm. This figure rather than that given above for accuracy represents the precision attainable in the measurement of small changes of resistance such as are usual in resistance thermometry.

WASHINGTON, October 1, 1914.